Harcourt
SCHOOL PUBLISHERS

Orlando Austin New York San Diego Toronto London

Visit *The Learning Site!*
www.harcourtschool.com

Heat Energy

Heat is energy that makes things hot. The sun is what gives Earth heat energy.

The sun is a star. It warms the land, air, and water all around you.

Fast Fact

The middle of the sun is at least 15 million degrees Celsius (27 million degrees Fahrenheit).

The sun gives heat energy to warm the sand, the water, and the air.

Some things warm up in the sunlight faster than other things. Dark-colored things warm up quickly. Light-colored things take longer to warm up.

CAUSE AND EFFECT **What happens when something is given heat energy?**

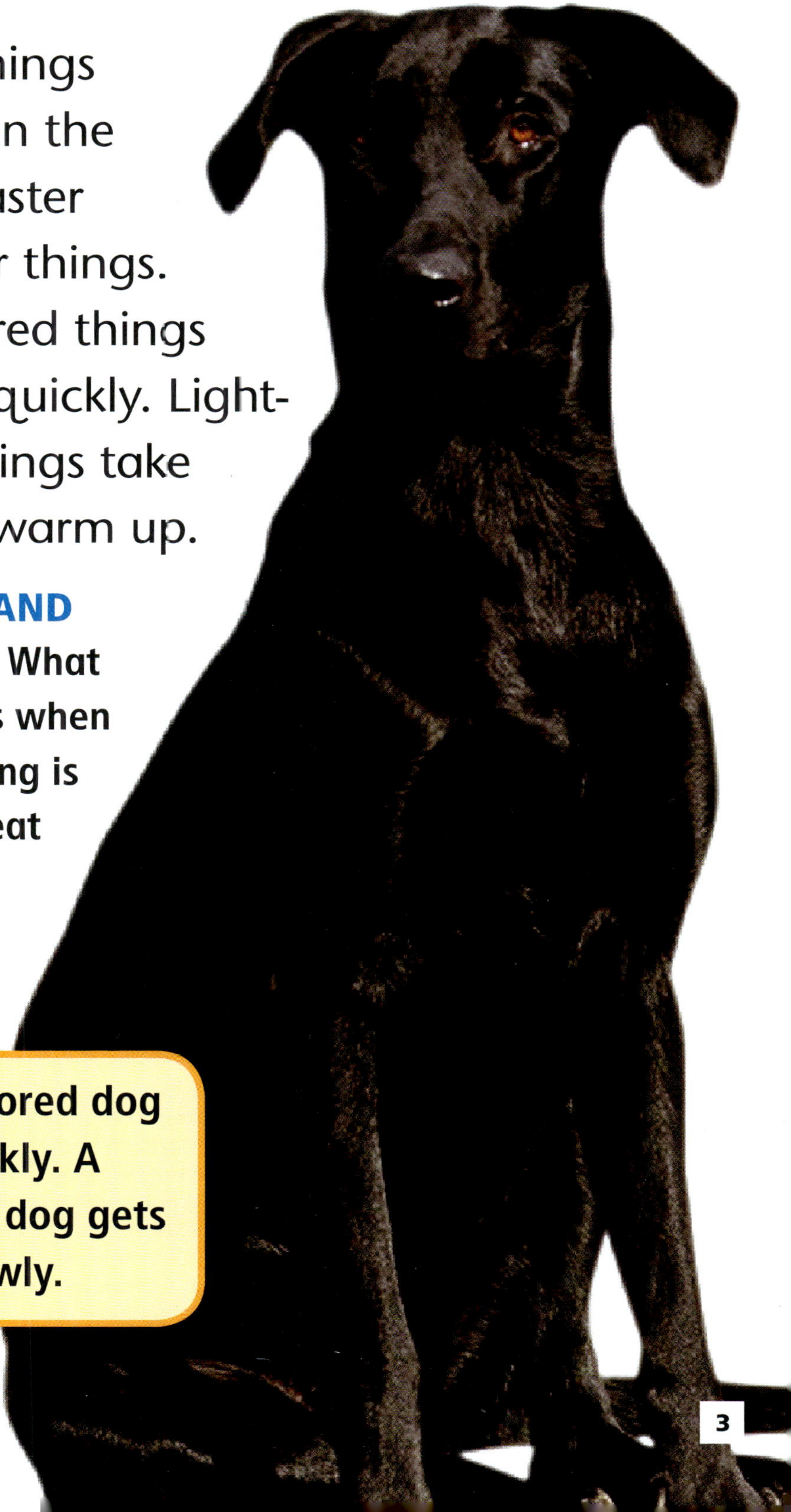

This dark-colored dog gets hot quickly. A light-colored dog gets hot more slowly.

Feeling the Heat

You feel heat from the sun. You can feel heat from other things, too.

Fire makes things warm. You can feel heat when something is burned.

Fire gives off heat. It warms this room.

You feel heat when your hands slip quickly against a rope.

Have you ever rubbed your hands together when they were cold? You can feel them get warm. Moving things together gives off heat.

Focus Skill **CAUSE AND EFFECT** **What things can you do to warm your hands?**

Light Energy

Light is a kind of energy. The sun gives light energy for the world around us.

When you go outside, you can see trees and houses. Light helps you to see.

The sun lights up the world around us.

A candle is burned to give light.

It can get dark at night. It can be hard to see. Fires and lamps give off light. People use fires and lamps to help them see.

MAIN IDEA AND DETAILS
Focus Skill
Where can we get light?

Fast Fact

A 100 watt light bulb at your house makes as much light as about 1,500 candles!

Light on the Move

Light moves. It passes through clear objects. It passes through clear glass. It passes through clear air.

These windows are made of clear glass. Light passes through them.

Objects that are not clear, block light. Trees, buildings, and even people can block light. They can make shadows. A **shadow** is a dark place made when an object blocks light. You can see lots of shadows on a sunny day.

Focus Skill **CAUSE AND EFFECT** What causes a tree to make a shadow?

Can you see shadows? What blocked the light from passing through?

Sound Energy

Did you know that sound is a kind of energy? **Sound** is energy you can hear. Music, talking, and a dog barking are all sound. Sound is all around us.

Musical instruments make many different sounds.

A fire engine gives light energy from its lights. It gives sound energy from its siren.

COMPARE AND CONTRAST **How is the light from the sun like sound from a drum?**

Summary

The sun sends heat and light energy to Earth. We also get heat and light in other ways. Light can pass through some things, but not others. Sound is a kind of energy, too.

Glossary

heat A kind of energy that makes things hot (2, 3, 4, 5, 11)

light A kind of energy that helps us see (6, 7, 8, 9, 11)

shadow A dark place made when an object blocks light (9)

sound A kind of energy that you can hear (10, 11)